The Book of Jubilees

Unveiling Universe Mystical Dimensions

Elan Nova

Copyright © 2024 by Elan Nova

All rights reserved. No part of this book may be reproduced, stored in a retrieval system, or transmitted in any form or by any means, electronic, mechanical, photocopying, recording, or otherwise, without prior written permission from the author, except for brief quotations in critical reviews or articles.

Table of Contents

Introduction — 5
    The Mysteries of Time and the Universe — 5
    Purpose of This Book — 6
Part I — 8
Foundations of the Jubilee — 8
Chapter 1 — 9
The Concept of Jubilee in Ancient Cultures — 9
    Origins and Historical Context — 10
    Symbolism of Cycles and Renewal — 12
Chapter 2 — 14
Sacred Time and Cosmic Rhythms — 14
    Time as a Divine Construct — 14
    Ancient Calendars and Cosmic Alignments — 15
    The Intersection of Time and the Cosmos — 18
Part II — 20
Mystical Dimensions of the Universe — 20
Chapter 3 — 21
Sacred Geometry and Universal Patterns — 21
    The Mathematics of Creation — 21
    Geometric Patterns in Nature and Beyond — 23
    The Spiritual Implications of Sacred Geometry — 25
Chapter 4 — 27
The Spiritual Fabric of Reality — 27
    Energy Fields and Vibrational Frequencies — 27
    Interconnectivity of All Things — 29
    The Spiritual Implications of the Fabric of Reality — 31

Part III	34
Insights from the Book of Jubilees	**34**
Chapter 5	35
Overview of the Book of Jubilees	**35**
Authorship, History, and Purpose	35
Key Themes and Messages	37
The Significance of the Book of Jubilees	39
Chapter 6	41
Chronology of Divine Events	**41**
Creation and Early History	41
The Role of Angels and Divine Laws	43
Theological Implications of Divine Chronology	45
Part IV	46
Unveiling the Mystical Dimensions	**46**
Chapter 7	46
The Interplay of the Celestial and Terrestrial	**47**
Influence of Heavenly Bodies	47
Portal Points and Sacred Spaces	49
The Spiritual Implications of the Celestial-Terrestrial Interplay	51
Chapter 8	53
Divine Wisdom and the Human Connection	**53**
Pathways to Enlightenment	53
Role of Prophecy and Revelation	55
The Connection Between Divine Wisdom, Prophecy, and Humanity	58
Part V	60
Reflections on Time and Redemption	60

Chapter 9	**61**
Jubilee Cycles and Human Evolution	**61**
Lessons from History	61
Prophetic Insights into Future Cycles	63
The Spiritual Significance of Jubilee Cycles for Human Evolution	65
Chapter 10	**68**
Redemption Through Sacred Knowledge	**68**
Embracing Mystical Teachings	68
Practical Applications for Modern Times	70
The Promise of Redemption Through Sacred Knowledge	73
Chapter 11	**75**
Timeline of Jubilees and Significant Events	**75**
Reflection on the Timeline	79
Conclusion	**80**
The Eternal Quest for Understanding	**80**
Closing Thoughts on the Jubilee Mysteries	81

Introduction

The Mysteries of Time and the Universe

The universe, in its vastness and complexity, is woven with threads of time, space, and energy. For centuries, philosophers, theologians, and scientists have sought to understand the intricate mechanisms that govern its operation. Time, often perceived as a linear journey from past to future, is far more enigmatic when viewed through a cosmic lens. Ancient civilizations, such as the Babylonians, Egyptians, and Mayans, regarded time as cyclical, reflecting the recurring rhythms of the natural world: the rising and setting of the sun, the phases of the moon, and the changing of seasons. These patterns were not mere coincidences but were seen as the heartbeat of the universe, evidence of a divine order underpinning all existence.

In the mystical traditions, time is not only a measure of sequential events but also a sacred force, a spiritual dimension that bridges the seen and unseen realms. It is believed to carry divine intent, shaping destinies and orchestrating cosmic events. The ancients observed that certain intervals of time—jubilees, cycles, and epochs—were imbued with unique energies that influenced human history, spiritual awakening, and cosmic transformation. These sacred periods were marked by renewal, restoration, and revelation, offering humanity opportunities to realign with higher truths.

Exploring the mysteries of the universe requires an understanding that reality is multi-dimensional. Modern physics, with its theories of relativity and quantum mechanics, has only begun to scratch the surface of these dimensions, echoing concepts that mystical traditions have held for millennia. Concepts like the interconnectedness of all things, the flow of energy through space, and the resonance of frequencies align with ancient teachings about the fabric of creation. The universe is not random; it is purposeful, alive, and charged with divine intelligence.

Purpose of This Book

This book, Unveiling Mystical Dimensions of the Universe, seeks to bridge the gap between ancient wisdom and modern understanding. At its heart lies the concept of the Jubilee, a sacred period in many traditions symbolizing liberation, restoration, and divine justice. The Jubilee is more than a historical observance; it is a lens through which we can decode the mysteries of existence, unlocking truths that transcend time and culture.

The purpose of this work is to provide readers with a comprehensive exploration of the mystical dimensions that govern the universe. While many texts focus on isolated aspects of cosmic mysteries, The book of Jubilee endeavors to present a holistic view. It will delve into sacred geometry, cosmic cycles, and the interplay between the celestial and terrestrial realms. Additionally, it will illuminate how ancient texts, like the Book of Jubilees, reveal profound insights about time, human destiny, and the eternal dance of creation.

This book is not merely an academic or speculative exercise; it is an invitation to embark on a journey of discovery and transformation. By understanding the divine patterns that shape our reality, readers will gain tools to navigate the complexities of modern life with greater clarity and purpose. The book of Jubilee aims to empower individuals to recognize their place in the cosmos and align their lives with the rhythms of the universe.

Ultimately, the book seeks to answer profound questions: **What is the purpose of time? How do celestial cycles influence our lives? What can ancient wisdom teach us about the nature of reality?** By addressing these inquiries, The Jubilee Chronicles aspires to not only inform but inspire, guiding readers toward a deeper connection with the mystical forces that sustain and shape the universe.

Part I

Foundations of the Jubilee

Chapter 1

The Concept of Jubilee in Ancient Cultures

The concept of the Jubilee originates in ancient times, deeply rooted in the spiritual and social structures of many early civilizations. The term "**Jubilee**" itself is derived from the Hebrew word **yovel**, which refers to the ram's horn blown to signify the commencement of this sacred time. In the Hebrew tradition, the Jubilee was established as a period of liberation and renewal, observed every fifty years as prescribed in the Torah (Leviticus 25). It was a time when debts were forgiven, slaves were freed, and land was returned to its original owners. The Jubilee represented a divine reset, a chance to realign society with principles of justice, equality, and divine order.

This concept, however, is not unique to the Hebrew tradition. Ancient cultures across the world held similar practices, each reflecting their understanding of cosmic cycles and the need for periodic renewal. The Mesopotamians, for example, practiced what was known as the "misharum" or "declaration of justice," during which debts were forgiven and social hierarchies were temporarily dissolved. The Egyptians celebrated the Heb-Sed festival, a ritual of renewal and restoration for the king, symbolizing the rejuvenation of his divine mandate to rule. Even in the Far East, ancient Chinese philosophies emphasized cycles of harmony and imbalance, leading to periods of societal and cosmic recalibration.

These practices demonstrate that the Jubilee is not merely a religious observance but a universal archetype, reflecting humanity's collective recognition of the need for balance, renewal, and the restoration of order. It reveals a profound understanding that time and history are cyclical rather than linear, requiring moments of pause to realign with the divine and natural laws.

Origins and Historical Context

The origins of the Jubilee concept can be traced to the **agrarian** societies of the ancient Near East, where the rhythms of life were dictated by the cycles of nature. In these early communities, survival depended on harmonious relationships—between individuals, within families, and with the land. Over time, these societies recognized the necessity of structured intervals to restore equilibrium, prevent exploitation, and safeguard communal well-being.

The Jubilee was formally codified in the Hebrew Bible, where it became a cornerstone of the Mosaic law. According to Leviticus 25, the Jubilee year followed seven cycles of seven years, marking the 50th year as a sacred time. The significance of the number seven, often associated with completion and divine perfection, underscored the spiritual essence of this observance. The Sabbatical year, observed every seven years, served as a precursor to the Jubilee, emphasizing rest for the land and a cessation of agricultural activity.

Historically, the Jubilee was more than a ritual; it was a **revolutionary concept**. In a world where power and wealth often became concentrated in the hands of a few, the Jubilee provided a system for redistributing resources and resetting social hierarchies. It aimed to prevent generational poverty, ensure access to land (the primary means of livelihood), and maintain social cohesion. This radical approach to justice and equity was a direct reflection of the divine character—merciful, just, and compassionate.

The cultural and historical importance of the Jubilee extended beyond the Hebrew context. In Babylonian society, for example, King Hammurabi and other rulers periodically declared debt cancellations to prevent economic disparity from destabilizing their kingdoms. Similarly, ancient Greek and Roman practices occasionally featured debt forgiveness, albeit less systematically than the Hebrew Jubilee. These examples highlight a shared understanding among ancient cultures: societies flourish when periodic renewal and redistribution are embraced.

Symbolism of Cycles and Renewal

At its core, the Jubilee embodies the profound symbolism of cycles and renewal, principles that are woven into the fabric of existence. Nature itself operates in cycles—day turns to night, seasons change, and life follows the pattern of birth, growth, death, and rebirth. The Jubilee mirrors these natural rhythms, reminding humanity of its interconnectedness with the cosmos and the divine.

The cycle of the Jubilee symbolizes **restoration**, both spiritual and material. It reflects the belief that all things originate from the Creator and must periodically return to their original state. This return is not merely about physical restitution but also spiritual reconciliation—restoring relationships, healing the land, and realigning human activities with divine will. For ancient cultures, this renewal was essential for maintaining cosmic harmony, ensuring that human actions did not disrupt the delicate balance of the universe.

The Jubilee also carries a deeper metaphysical significance. In its cycles, it reveals the divine pattern of time—ordered, purposeful, and redemptive. **It teaches that time is not a relentless march toward entropy but a sacred spiral, offering opportunities for growth, reflection, and transformation**. Each Jubilee marks the completion of a cycle and the beginning of a new one, symbolizing hope, renewal, and the promise of divine intervention.

Furthermore, the Jubilee serves as a reminder of the **impermanence of human possessions and achievements.** By returning land to its original owners and canceling debts, the Jubilee reaffirms the principle that everything ultimately belongs to the Creator. This act of surrender fosters humility, gratitude, and a sense of stewardship rather than ownership.

The symbolism of cycles and renewal embedded in the Jubilee extends beyond the physical and societal realms to the spiritual journey of individuals. **It calls for self-examination, repentance, and a recommitment to living in harmony with divine principles.** It is a time to let go of burdens, forgive grievances, and embrace new beginnings.

In summary, the Jubilee is a timeless concept, reflecting humanity's intrinsic need for balance, renewal, and alignment with cosmic and divine laws. Its origins and practices, deeply rooted in ancient cultures, reveal universal truths about the nature of existence and the rhythms that govern our lives.

Chapter 2

Sacred Time and Cosmic Rhythms

Time as a Divine Construct

Time, as perceived in ancient traditions, was far more than a chronological sequence of events. It was regarded as a divine construct, intricately designed by a higher power to govern the rhythms of the universe and the lives of its inhabitants. Unlike the modern, linear understanding of time, ancient cultures viewed it as cyclical, repeating in patterns that reflected the natural order of creation. This cyclical view acknowledged time as sacred, infused with spiritual significance, and deeply connected to the cosmic and divine.

The notion of time as sacred can be seen in the way it was integrated into religious practices and societal structures. In the Hebrew tradition, for example, the Sabbath was a weekly marker of sacred time, set apart for rest and worship, mirroring the divine rest on the seventh day of creation. Similarly, the Jubilee year, as outlined in Leviticus 25, was another sacred time, emphasizing renewal, liberation, and the restoration of harmony. These observances were not arbitrary but reflected a belief that time itself carried divine intent, shaping human lives and the world in alignment with God's purposes.

In many ancient cultures, time was also seen as a tool for divine communication. Certain moments in time were considered **"kairos"**—an opportune, spiritually significant time—rather than "chronos," the quantitative measurement of time. This distinction underscored the idea that time could be a medium for divine intervention, where events of great cosmic and spiritual importance unfolded. Festivals, rituals, and sacred observances were often aligned with these significant moments, further reinforcing the concept of time as a bridge between the earthly and the divine.

This sacred understanding of time also influenced human behavior and societal organization. People lived in tune with the rhythms of the universe, aligning their actions with the cycles of nature and divine commands. Time, in this sense, was not a commodity to be managed but a sacred gift to be revered and harmonized with.

Ancient Calendars and Cosmic Alignments

Ancient civilizations, keenly aware of the sacredness of time, developed sophisticated calendars to track its passage and align human activities with cosmic rhythms. These calendars were not merely tools for measuring days and months; they were profound systems that reflected a deep understanding of the universe's order and humanity's place within it.

The Hebrew calendar, for instance, was rooted in a lunisolar system, combining the cycles of the moon and the sun to mark sacred times. The alignment of the calendar with agricultural seasons and religious observances highlighted the interconnectedness of human life, natural cycles, and divine mandates. Festivals such as Passover, Sukkot, and the Jubilee year were carefully timed to coincide with specific phases of the moon or solar cycles, symbolizing renewal, harvest, and liberation.

Similarly, **the Mayan calendar** exemplifies the ancient understanding of time as a cosmic rhythm. The Mayans created multiple interlocking cycles, including the Tzolk'in (a 260-day ceremonial calendar) and the Haab' (a 365-day solar calendar), which together formed a larger cycle known as the Calendar Round. Beyond their practical use, these calendars were deeply spiritual, reflecting the Mayan belief in the cyclical nature of time and its role in cosmic harmony. The Long Count calendar, famously associated with the "end of the world" interpretations in 2012, was a means of tracking vast cosmic cycles, demonstrating the Mayans' perception of time as eternal and recurring.

In Egypt, the civil calendar was complemented by the Sothic cycle, based on the heliacal rising of the star Sirius, which marked the annual flooding of the Nile.

This event was not only crucial for agriculture but also held spiritual significance, symbolizing renewal and the sustenance of life. The Egyptians' precise astronomical observations and their integration into their calendar system reveal their sophisticated understanding of the relationship between celestial movements and earthly events.

The Babylonians, another advanced civilization, also developed intricate calendar systems, integrating lunar cycles with solar observations. Their calendar was instrumental in their religious practices and agricultural planning. They meticulously tracked celestial bodies, associating their movements with divine messages and omens. This fusion of astronomy and theology underscores their belief that time was governed by divine forces and that humanity's actions must align with these celestial rhythms.

In ancient China, the traditional lunar-solar calendar reflected the harmony between heaven and earth. The calendar was used to mark auspicious days for agricultural activities, rituals, and festivals, emphasizing the cyclical nature of time and its relationship to human endeavors. The Chinese concept of Tian (heaven) as a guiding force in time and space reinforced the idea that cosmic order dictated earthly affairs.

These ancient calendars demonstrate a universal recognition of cosmic rhythms and their influence on human life. The alignment of sacred observances with celestial phenomena reflected a profound understanding of the interconnectedness between the heavens and the earth. By synchronizing their lives with these rhythms, ancient peoples sought to maintain harmony with the divine order, ensuring balance and prosperity in their communities.

The Intersection of Time and the Cosmos

The creation of ancient calendars was rooted in a meticulous observation of celestial bodies—the sun, moon, stars, and planets—and their movements across the sky. These observations were not merely scientific but deeply spiritual, as they revealed patterns that mirrored the sacred rhythms of time.

For example, the lunar cycle, with its phases of waxing and waning, symbolized life's transitions and the interplay of light and darkness. The solar cycle, with its solstices and equinoxes, represented the balance of opposites—day and night, growth and rest. Together, these cycles created a framework for understanding time as a dynamic interplay of forces, each contributing to the greater whole.

The study of celestial alignments also extended to the architecture and planning of ancient monuments. Structures such as Stonehenge in England, the Great Pyramids of Giza in Egypt, and the observatories of the Mayans were designed to align with specific astronomical events. These alignments were not coincidental but intentional, reflecting the belief that sacred sites served as portals between the earthly and celestial realms. The positioning of these structures allowed ancient peoples to track time, mark sacred moments, and engage in rituals that connected them to the divine.

In essence, ancient calendars and cosmic alignments were tools for living in harmony with the universe's sacred rhythms. They provided a framework for understanding time as a divine construct, revealing its cyclical, purposeful nature. By studying and honoring these rhythms, ancient cultures created systems that bridged the physical and spiritual, ensuring that their lives remained aligned with the eternal patterns of the cosmos. This profound integration of time, spirituality, and cosmic observation remains a testament to humanity's enduring quest to understand its place in the universe.

Part II

Mystical Dimensions of the Universe

Chapter 3

Sacred Geometry and Universal Patterns

Sacred geometry, often referred to as the "language of creation," is the study of shapes, patterns, and proportions that are fundamental to the design of the universe. Across ancient cultures and modern science, sacred geometry reveals the interconnectedness of all things, demonstrating how universal principles manifest through mathematical and geometric order. From the spiraling galaxies in the cosmos to the structure of DNA, sacred geometry provides a framework for understanding the harmony and beauty of existence.

The Mathematics of Creation

The universe operates on precise mathematical principles, often described as the building blocks of creation. These principles govern the structures of physical reality and connect everything in a cohesive, harmonious system. Let's go into the key aspects that illuminate the role of mathematics in the creation of the universe:

1. The Golden Ratio (Phi)
   - The Golden Ratio, approximately 1.618, is a mathematical constant found in the proportions of

natural and human-made structures. It represents an ideal of balance and harmony.

- In nature, Phi appears in the arrangement of leaves on a stem, the spiral of seashells, and the branching of trees. Artists and architects, including Leonardo da Vinci, have used it to create works of art and buildings that are aesthetically pleasing.

2. The Fibonacci Sequence

- The Fibonacci Sequence is a series of numbers where each number is the sum of the two preceding ones (e.g., 0, 1, 1, 2, 3, 5, 8, 13).

- This sequence mirrors the growth patterns found in nature, such as the arrangement of sunflower seeds, the spiral of a nautilus shell, and the scales of pinecones. It is a mathematical expression of how life unfolds in a harmonious and predictable manner.

3. Pi (π) and the Circle

- Pi, the ratio of a circle's circumference to its diameter, is a universal constant appearing in geometry, physics, and cosmology.

- The circle itself is a sacred symbol, representing wholeness, infinity, and cycles. It is foundational to celestial movements, such as the orbits of planets and the paths of stars.

4. Fractals: Infinite Patterns

- Fractals are complex geometric patterns that repeat at every scale. They are evident in snowflakes, river networks, and even human lungs.

- Mathematically, fractals demonstrate how the infinite complexity of the universe can arise from simple rules. They reveal a self-organizing principle that is both intricate and ordered.

Geometric Patterns in Nature and Beyond

The geometry of the natural world reflects the underlying principles of sacred design. These patterns are not random but are expressions of the same universal truths that govern the cosmos. Below are examples of geometric patterns observed in nature and their connection to the mystical dimensions of the universe:

1. Spirals

- Spirals, such as those seen in galaxies, hurricanes, and shells, are governed by the Fibonacci Sequence and the Golden Ratio.
- These forms symbolize growth, expansion, and the continuity of life, reflecting the universe's dynamic nature.

2. Hexagons

- The hexagon is a prevalent shape in nature, evident in honeycombs and the crystalline structure of snowflakes.
- It represents efficiency and interconnectivity, as it is one of the most efficient shapes for packing and structuring matter. In mysticism, it symbolizes balance and unity.

3. The Flower of Life
- The Flower of Life is a geometric design made up of overlapping circles. It has been found in ancient Egyptian temples, Indian mandalas, and other cultural artifacts.
- This symbol represents the interconnectedness of all life and the blueprint of creation. Its patterns contain the basis for other sacred geometric forms, such as the Seed of Life and the Tree of Life.

4. Platonic Solids
- The five Platonic Solids (tetrahedron, cube, octahedron, dodecahedron, and icosahedron) are three-dimensional shapes with identical faces, edges, and angles.
- These solids are associated with the classical elements (fire, earth, air, water, and ether) and are thought to represent the fundamental building blocks of reality.

5. The Spiral of the Cosmos
- Spiral galaxies, including the Milky Way, mirror the same mathematical principles that govern smaller spirals on Earth.
- This repetition across scales, from the microscopic to the cosmic, illustrates the fractal nature of the universe, where the same patterns are echoed at every level of creation.

The Spiritual Implications of Sacred Geometry

Sacred geometry is not merely a scientific observation; it is a spiritual understanding of how divine intelligence is embedded in the structure of the universe. These patterns and proportions serve as a reminder of the interconnectedness of all life and the harmony that underlies creation. By studying sacred geometry, one gains insight into the deeper truths of existence and the divine order that governs all things.

1. Unity and Interconnection
   - The recurring patterns found in sacred geometry underscore the oneness of creation. Every structure, from atoms to galaxies, is part of a unified whole.

2. Harmony and Balance
   - The principles of sacred geometry reflect a universe that is inherently balanced and harmonious. Understanding these principles can inspire individuals to seek balance in their own lives.

3. Pathways to Understanding the Divine
   - Many mystical traditions view sacred geometry as a pathway to connect with the divine. It is considered a key to unlocking the mysteries of existence, enabling one to see the divine blueprint in all things.

Sacred geometry and universal patterns are the mathematical fingerprints of creation, demonstrating the harmony, balance, and divine intelligence that govern the cosmos. By exploring these principles, we uncover not only the physical structure of the universe but also its spiritual dimensions, offering a profound perspective on the interconnectedness and purpose of all existence.

Chapter 4

The Spiritual Fabric of Reality

The spiritual fabric of reality refers to the underlying essence of existence, a complex web of energy, frequencies, and interconnectedness that unites all living and non-living things. This concept transcends the material world, suggesting that reality is not solely physical but also deeply spiritual, woven together by unseen forces and energies. Ancient traditions, modern science, and spiritual philosophies converge to reveal a universe imbued with divine intelligence, infinite connectivity, and purposeful design.

Energy Fields and Vibrational Frequencies

At the heart of the spiritual fabric of reality lies the understanding that everything in the universe is composed of energy. From the smallest subatomic particles to vast galaxies, all matter vibrates at specific frequencies, creating an intricate interplay of energy fields. These vibrational frequencies are not just physical phenomena but also carry spiritual and metaphysical significance, shaping the way reality is perceived and experienced.

1. The Quantum Perspective on Energy
- In modern physics, quantum mechanics has revealed that matter is not solid but consists of particles vibrating at different frequencies. These particles are connected by fields of energy, creating a dynamic and ever-changing reality.
- This scientific understanding parallels ancient spiritual teachings that describe the universe as a field of energy in constant motion, guided by divine forces.

2. The Human Energy Field
- Human beings are not merely physical entities but energetic ones. Ancient practices like Ayurveda, Traditional Chinese Medicine, and Reiki emphasize the presence of an energy field surrounding and flowing through the body. This field, often called the aura, is composed of layers that reflect physical, emotional, mental, and spiritual health.
- The body's vibrational frequency can be influenced by thoughts, emotions, and external factors. Positive energy and high vibrations are associated with health, joy, and spiritual alignment, while low vibrations often correlate with illness, negativity, and imbalance.

3. The Resonance of Frequencies
- Just as musical instruments resonate with certain frequencies, humans and the environment are constantly interacting and influencing each other's vibrations. This phenomenon, known as entrainment, explains how one energy field can align with another.

- For example, the vibrations of the earth (Schumann Resonance) and the frequencies of the human brain can synchronize, creating a harmonious connection that promotes well-being and spiritual clarity.

4. Manifestation and Vibrational Energy
- Many spiritual traditions teach that thoughts and emotions emit vibrational frequencies, which shape reality. This principle is central to practices like the Law of Attraction, which suggests that positive vibrations attract positive outcomes, and negative vibrations attract challenges.
- The idea of vibrational alignment highlights the power of mindfulness, gratitude, and meditation in raising one's frequency and transforming life experiences.

Interconnectivity of All Things

The interconnectedness of all things is a foundational principle of the spiritual fabric of reality. This concept asserts that no entity exists in isolation; rather, everything in the universe is interwoven through an invisible web of energy, consciousness, and purpose. From ecosystems on Earth to the movements of galaxies, all aspects of creation are deeply linked, forming a cohesive and harmonious whole.

1. The Web of Life
- Indigenous traditions often describe the universe as a "web of life," emphasizing the interconnectedness of

humans, animals, plants, and the natural world. This perspective teaches respect and reverence for all forms of life, recognizing that every action has consequences that ripple through the fabric of existence.

- Scientific studies in ecology support this idea, demonstrating how ecosystems are delicate balances of interconnected organisms. Disruption in one part of the system inevitably impacts the whole.

2. Collective Consciousness

- The concept of collective consciousness suggests that all beings are connected through a shared spiritual awareness. This universal consciousness transcends individual experience, uniting humanity and all life in a greater, divine intelligence.

- Practices like prayer, meditation, and communal rituals often tap into this collective field, amplifying spiritual energy and fostering unity.

3. Sacred Geometry and Interconnection

- Patterns found in sacred geometry, such as the Flower of Life, illustrate the interconnectedness of all things. These geometric forms are considered the blueprint of creation, showing how individual components fit together to form a unified whole.

- This interconnectedness is mirrored in fractals, where the same patterns repeat across different scales, symbolizing the infinite connectivity of the universe.

4. Universal Energy Exchange

- The interconnectivity of the universe is facilitated by a constant exchange of energy. Whether through the transfer of nutrients in ecosystems, the flow of electromagnetic waves, or the interactions between human energy fields, this exchange sustains life and maintains balance.
- Spiritual traditions often encourage practices that honor this exchange, such as acts of kindness, gratitude, and giving, which foster harmony and align with the natural flow of the universe.

5. The Butterfly Effect in Spirituality

- The interconnected nature of the universe is beautifully illustrated by the "butterfly effect," a concept from chaos theory that suggests small actions can have far-reaching consequences.
- Spiritually, this principle reinforces the idea that even the smallest thoughts, words, or deeds contribute to the larger reality, encouraging mindfulness and intentionality.

The Spiritual Implications of the Fabric of Reality

Understanding the spiritual fabric of reality has profound implications for how we perceive ourselves, others, and the universe. It shifts the focus from separation to unity, revealing that every individual is part of a greater whole. By aligning with the energy fields and vibrations of the universe, individuals can experience deeper harmony, purpose, and spiritual growth.

1. Unity and Oneness
- Recognizing the interconnectedness of all things fosters a sense of unity and compassion. It helps individuals see themselves as part of a larger cosmic family, transcending divisions based on race, culture, or belief.

2. Healing and Balance
- Practices like meditation, energy healing, and sound therapy leverage vibrational frequencies to restore balance and promote well-being. By aligning with higher frequencies, individuals can heal not only themselves but also their environments.

3. Living in Harmony with the Universe
- The spiritual fabric of reality calls for a life of alignment with natural rhythms and divine principles. It encourages practices like mindfulness, sustainable living, and acts of kindness, which honor the interconnectedness of existence.

4. Awakening to Purpose
- Understanding that every action, thought, and emotion contributes to the whole inspires individuals to live with intention and purpose. It transforms daily life into a sacred journey of co-creating reality with the divine.

The spiritual fabric of reality weaves together energy fields, vibrations, and the interconnectedness of all things, revealing a universe that is deeply harmonious, purposeful, and alive. By understanding and aligning with this fabric, individuals can transcend the illusion of separation, embrace unity, and connect with the divine essence that permeates all existence.

Part III

Insights from the Book of Jubilees

Chapter 5

Overview of the Book of Jubilees

The Book of Jubilees, also known as "The Little Genesis," is one of the most intriguing ancient texts that offers a unique retelling of biblical history. It expands on stories found in the Book of Genesis and parts of Exodus, presenting them within a structured framework of jubilees—periods of 49 years divided into seven cycles of seven years. The text is deeply rooted in Jewish tradition, providing insights into the sacred understanding of time, divine law, and human destiny. Though not part of the canonical scriptures for most traditions, it has significant historical, theological, and mystical value.

Authorship, History, and Purpose

1. Authorship

   - The Book of Jubilees is believed to have been written in the 2nd century BCE, during the Second Temple period. While the exact author is unknown, it is widely accepted that the text was composed by a Jewish scholar or priest deeply versed in the Torah.
   - The text was likely written in Hebrew, though much of it survives in Ge'ez (the liturgical language of Ethiopia),

fragments of Hebrew found among the Dead Sea Scrolls, and translations in Greek and Latin.
- Its author demonstrates an intense focus on the laws of Moses and portrays a worldview that emphasizes the importance of divine order, sacred time, and covenantal relationships.

2. Historical Context

- The Book of Jubilees emerged during a turbulent time in Jewish history, marked by the rise of Hellenistic influences and the struggle to preserve Jewish identity and traditions.
- The text reflects a response to these cultural pressures, aiming to reinforce Jewish laws and the sanctity of the Torah. It reinterprets the Genesis narrative to emphasize obedience to divine law and the importance of observing sacred times.
- Its inclusion of angelic mediation and the heavenly tablets suggests an attempt to present the Torah not merely as a human document but as a divine revelation rooted in cosmic order.

3. Purpose

- The primary purpose of the Book of Jubilees is to offer a sacred chronology that aligns human history with divine will. By dividing history into jubilees, the text underscores the cyclical nature of time and its connection to divine patterns.
- The book aims to strengthen the Jewish community's commitment to the covenant with God, particularly

through adherence to the Sabbath, festivals, and dietary laws.

- It also serves as a guide to moral and ethical living, emphasizing themes of repentance, obedience, and the ultimate restoration of creation to its intended harmony.

Key Themes and Messages

1. Sacred Time and Divine Order

- A central theme of the Book of Jubilees is the concept of sacred time, illustrated by its jubilee structure. This framework reinforces the idea that time itself is a divine construct, governed by specific laws and cycles.

- The text repeatedly emphasizes the sanctity of the Sabbath, presenting it as a cornerstone of the covenant between God and humanity. Observance of sacred times, including festivals, is portrayed as essential for maintaining alignment with divine will.

2. Covenant and Obedience

- The Book of Jubilees places significant emphasis on the covenant established between God and the patriarchs, particularly Abraham, Isaac, and Jacob.

- It highlights the importance of obedience to God's commandments as a means of upholding this covenant. The text often presents the patriarchs as models of perfect obedience, contrasting them with those who stray from divine law.

3. Heavenly Tablets and Predestination

- A unique aspect of the Book of Jubilees is its reference to the "heavenly tablets," which are said to contain a divine record of all events—past, present, and future. This idea underscores the belief in a preordained plan for creation, where every event unfolds according to God's design.

- The heavenly tablets symbolize the alignment between earthly actions and heavenly decrees, emphasizing the importance of living in harmony with divine law.

4. Moral Lessons and Ethical Living

- The text is rich in moral teachings, using the stories of biblical characters to impart lessons about virtue, repentance, and the consequences of sin. For example, it expands on the story of Cain and Abel to emphasize the dangers of jealousy and the sanctity of life.

- The narrative often portrays the patriarchs as paragons of righteousness, offering readers examples of ethical living that align with divine expectations.

5. Angelic Mediation and Spiritual Realms

- The Book of Jubilees frequently mentions the role of angels in mediating between God and humanity. Angels are depicted as messengers, record-keepers, and enforcers of divine will.

- This focus on angelic beings highlights the interconnectedness between the earthly and heavenly realms, suggesting that human actions have spiritual consequences.

6. Restoration and Redemption
- Another key message of the Book of Jubilees is the promise of restoration. It envisions a time when creation will be restored to its original harmony, free from sin and corruption.
- This theme reflects the hope for redemption and the ultimate fulfillment of God's plan for humanity and the cosmos.

The Significance of the Book of Jubilees

The Book of Jubilees is more than a retelling of biblical stories; it is a profound exploration of time, law, and divine purpose. Its unique perspective offers valuable insights into Jewish thought during the Second Temple period, shedding light on how sacred texts were interpreted and expanded to address contemporary challenges.

For modern readers, the Book of Jubilees serves as a reminder of the cyclical nature of time, the importance of living in alignment with divine principles, and the interconnectedness of earthly and spiritual realities. Its messages about covenant, obedience, and restoration remain timeless, offering a path to deeper understanding and spiritual growth.

By delving into its themes and historical context, one can appreciate the Book of Jubilees not only as a historical document but also as a source of profound spiritual wisdom. It bridges the human and the divine, inviting readers to see their lives as part of a larger, sacred narrative that spans the cosmos.

Chapter 6

Chronology of Divine Events

The Book of Jubilees presents a meticulously detailed chronology of divine events, offering a sacred retelling of biblical history framed within the structure of jubilees. This approach highlights the significance of time as a divine construct and underscores the intentionality of God's actions throughout history. By organizing events into cycles of 49 years, the text connects human history with cosmic rhythms, portraying every milestone as part of a preordained plan.

Creation and Early History

The Book of Jubilees begins with the creation narrative, aligning closely with the account in Genesis but adding layers of interpretation that emphasize sacred time and divine order. The text portrays creation as an act of deliberate design, carried out in harmony with the celestial realms. God is depicted as establishing the heavens, the earth, and all living beings within the framework of seven days, a sacred pattern that recurs throughout the book.

1. The Creation of Time and Order

- The first act of creation in the Book of Jubilees is the establishment of time itself. Days, months, and years are divinely ordained, setting the stage for the cycles that govern the universe.
- The calendar system outlined in the text reflects a deep understanding of celestial movements, emphasizing the sanctity of the Sabbath and the importance of aligning human activity with divine rhythms.

2. Adam and Eve: The First Covenant

- The story of Adam and Eve is expanded in the Book of Jubilees, emphasizing their role as the first recipients of God's covenant. They are instructed to observe the Sabbath, symbolizing humanity's duty to honor sacred time.
- Their expulsion from Eden is portrayed not only as a consequence of disobedience but also as a disruption of divine harmony, setting the stage for humanity's ongoing struggle to realign with God's will.

3. The Generations of Early Humanity

- The text chronicles the descendants of Adam and Eve, highlighting key figures such as Seth, Enosh, and Noah. These individuals are portrayed as custodians of divine wisdom, tasked with preserving God's laws and passing them on to future generations.
- The narrative also explores the moral decline of humanity, culminating in the flood, which is presented as both a judgment and a renewal of creation.

The Role of Angels and Divine Laws

Angels play a central role in the Book of Jubilees, serving as mediators between God and humanity, guardians of divine laws, and participants in key historical events. The text emphasizes the interconnectedness of the heavenly and earthly realms, portraying angels as instruments of divine will who ensure that God's plan unfolds according to the heavenly tablets.

1. Angelic Participation in Creation
   - The Book of Jubilees attributes specific roles to angels during creation, depicting them as witnesses to God's work and as stewards of cosmic order. Angels are responsible for maintaining the harmony of the universe, reflecting their integral role in the divine hierarchy.

2. Angels as Teachers and Guides
   - The text frequently highlights the role of angels in imparting divine knowledge to humanity. For instance, the angel of the presence narrates the entire Book of Jubilees to Moses, providing him with a comprehensive understanding of history and divine law.
   - This act underscores the belief that sacred knowledge originates from the heavenly realms and must be faithfully transmitted to humanity.

3. The Watchers and the Fall of Angels
   - One of the most striking narratives in the Book of Jubilees is the story of the Watchers, a group of angels

who descended to earth and corrupted humanity by teaching forbidden knowledge. Their actions led to widespread sin and disorder, ultimately resulting in the flood.

- This account emphasizes the dangers of deviating from divine laws and the consequences of disrupting the balance between the heavenly and earthly realms.

4. Angels as Guardians of Divine Law

- Angels are portrayed as custodians of God's commandments, ensuring that humanity adheres to the divine covenant. They are depicted as recording human actions in the heavenly tablets, reinforcing the idea that every deed has eternal significance.

- The text also describes angels as enforcers of divine justice, carrying out God's judgments on individuals and nations who defy His laws.

5. The Interplay Between Angels and Human Destiny

- The Book of Jubilees portrays angels as deeply involved in shaping human destiny, guiding individuals and communities toward fulfillment of God's plan. Their interventions are often subtle yet profound, reflecting the interconnectedness of the spiritual and material worlds.

- This dynamic highlights the importance of aligning human actions with divine intent, as angels work to ensure that humanity remains on the path of righteousness.

Theological Implications of Divine Chronology

The Book of Jubilees weaves together the narratives of creation, early history, and angelic involvement to present a cohesive vision of divine chronology. It emphasizes that every event in history, no matter how small, is part of a greater cosmic plan. This perspective invites readers to view their own lives as interconnected with this sacred timeline, encouraging them to live in alignment with divine laws and rhythms.

The chronology of divine events in the Book of Jubilees offers a profound understanding of God's plan for humanity and the universe. By exploring creation, early history, and the role of angels, the text provides a comprehensive framework for understanding the interplay between time, law, and divine purpose. Through this lens, the reader is invited to see the world as a sacred tapestry, where every thread contributes to the unfolding of a divine masterpiece.

Part IV

Unveiling the Mystical Dimensions

Chapter 7

The Interplay of the Celestial and Terrestrial

The universe operates as a dynamic system where the celestial and terrestrial realms are intricately connected. Ancient wisdom and modern science alike affirm that heavenly bodies influence life on Earth, while certain terrestrial locations act as portals or sacred spaces where this connection is heightened. The interplay between the celestial and terrestrial is not merely physical but deeply spiritual, influencing human existence, natural phenomena, and the unfolding of cosmic events.

Influence of Heavenly Bodies

1. The Role of the Sun and Moon

   - The sun and moon are perhaps the most evident celestial influences on Earth. The sun governs the cycles of day and night, driving photosynthesis and life processes, while the moon affects tides, biological rhythms, and emotional states.

   - In spiritual traditions, the sun often symbolizes divine illumination, vitality, and masculine energy, whereas the moon represents intuition, cycles, and feminine energy. This duality reflects the balance between light and darkness, action and reflection.

2. Planetary Alignments and Their Effects

- Planets have long been believed to influence human behavior and events on Earth. Astrology, an ancient system of knowledge, maps the positions of planets at specific times, interpreting their alignment as having an impact on individual destinies and collective energies.
- For example, Mercury retrograde is associated with disruptions in communication and technology, while Venus's position is linked to matters of love and beauty. Whether viewed scientifically or spiritually, these planetary movements reflect the interconnectedness of celestial mechanics and human life.

3. The Zodiac and Cosmic Archetypes

- The twelve zodiac signs correspond to constellations in the heavens, serving as archetypes that influence human characteristics and spiritual paths. Each sign is ruled by specific celestial bodies, reinforcing the idea that the heavens shape earthly life.
- This relationship reflects the belief that the cosmos mirrors the inner workings of the human spirit, a concept often summarized as "As above, so below."

4. Seasonal and Celestial Cycles

- The movement of the Earth around the sun creates the seasons, which have profound implications for agriculture, culture, and spirituality. Equinoxes and solstices mark significant celestial events celebrated in many traditions as times of renewal and reflection.

- These cycles are seen as opportunities to align human activities with cosmic rhythms, ensuring harmony with the greater universe.

5. The Influence of Stars and Constellations
- Stars and constellations have served as navigational guides and spiritual symbols for millennia. The North Star, for example, has been a beacon of guidance, while the constellation Orion often represents strength and cosmic connection in various cultures.
- The alignment of stars with certain terrestrial monuments, such as the Pyramids of Giza, suggests that ancient civilizations understood and honored the profound relationship between the heavens and the Earth.

Portal Points and Sacred Spaces

Sacred spaces and portal points are terrestrial locations believed to serve as conduits between the earthly and celestial realms. These places are often associated with heightened spiritual energy, allowing for deeper connection with the divine and the cosmos.

1. Ley Lines and Energy Grids
- Ley lines are hypothetical alignments of landmarks, such as ancient monuments, megaliths, and natural formations, believed to form an energy grid across the Earth. These lines are thought to channel cosmic energy, creating powerful intersections known as vortex points.

- Sacred sites like Stonehenge, Machu Picchu, and the Great Pyramids are often situated on these lines, suggesting that ancient builders intentionally aligned their structures with this energy network.

2. Mountains as Sacred Portals

- Mountains are often regarded as sacred spaces, symbolizing a bridge between Earth and heaven. In many cultures, they are considered dwelling places of gods or locations where divine revelation occurs.
- For example, Mount Sinai is central to the biblical narrative as the site where Moses received the Ten Commandments, while Mount Kailash in Tibet is revered as a cosmic axis connecting the physical and spiritual worlds.

3. Temples and Monuments Aligned with Celestial Events

- Many ancient structures are aligned with celestial events, underscoring the interplay between the terrestrial and celestial. The Temple of Karnak in Egypt aligns with the solstice sun, while Chichen Itza in Mexico features a serpent-like shadow during the equinoxes.
- These alignments suggest that sacred architecture was designed to harness celestial energy, allowing worshippers to connect with cosmic forces during significant times.

4. Natural Portals: Lakes, Forests, and Caves

- Certain natural formations, such as lakes, forests, and caves, are also considered sacred spaces. These areas are often associated with myths of creation or divine intervention.

- For example, the Oracle of Delphi in ancient Greece operated within a sacred cave, believed to channel prophetic energy from the gods.

5. The Role of Rituals in Sacred Spaces
- Rituals performed in sacred spaces amplify the connection between the celestial and terrestrial realms. These practices, such as offerings, meditations, and prayers, are believed to open portals of communication, allowing divine energies to flow freely.
- The use of specific symbols, sounds, and materials in these rituals aligns human intention with cosmic forces, creating a harmonious exchange between dimensions.

The Spiritual Implications of the Celestial-Terrestrial Interplay

The relationship between the celestial and terrestrial realms reveals the universe as a unified whole, where the macrocosm of the heavens reflects the microcosm of earthly life. This interconnectedness carries profound spiritual implications:

1. Alignment with Cosmic Rhythms
- By understanding and aligning with celestial movements, individuals can harmonize their lives with the greater universe. Practices like observing the

phases of the moon or celebrating solstices allow for deeper spiritual awareness and balance.

2. Sacred Responsibility
- Recognizing the influence of celestial bodies and sacred spaces emphasizes humanity's role as stewards of both the Earth and its spiritual heritage. It encourages reverence for natural and cultural landmarks that serve as connections to the divine.

3. Enhanced Spiritual Awareness
- The understanding of portal points and celestial influences inspires individuals to seek deeper connections with the universe. Visiting sacred spaces or meditating on celestial movements can heighten spiritual perception and foster personal growth.

The interplay between the celestial and terrestrial is a testament to the interconnectedness of all creation. From the influence of heavenly bodies to the sacred energy of terrestrial spaces, this relationship invites humanity to see the universe as a harmonious, living system. By understanding and honoring these connections, individuals can align with the cosmos, tapping into the spiritual dimensions that sustain and guide life.

Chapter 8

Divine Wisdom and the Human Connection

The universe operates as a vast and intricate system governed by divine wisdom—a higher intelligence that orders creation, sustains life, and guides humanity toward its ultimate purpose. This wisdom, often described as sacred or cosmic knowledge, permeates every aspect of existence. Humans, as part of this divine design, are endowed with the ability to connect with this wisdom, uncovering profound truths that transcend the material world.

Pathways to Enlightenment

Enlightenment, in spiritual terms, refers to a state of heightened awareness and unity with divine truth. Achieving this state requires aligning oneself with divine wisdom and embracing practices that nurture spiritual growth. The Book of Jubilees and other ancient texts suggest several pathways through which humans can access divine wisdom and move toward enlightenment.

1. Contemplation and Meditation
   - The practice of meditation has long been regarded as a gateway to enlightenment. By stilling the mind and focusing inward, individuals can transcend the

distractions of the material world and attune themselves to the divine.

- In many traditions, meditation involves contemplation of sacred texts, prayers, or the mysteries of the universe, allowing individuals to receive insights that lead to a deeper understanding of their purpose.

2. Adherence to Divine Laws

- Observing divine commandments is another key pathway to enlightenment. The Book of Jubilees emphasizes that living in alignment with God's laws fosters harmony with the universe and opens channels to divine wisdom.

- Practices such as honoring the Sabbath, observing sacred times, and leading a life of moral integrity help individuals cultivate spiritual awareness and connect with higher truths.

3. Sacred Rituals and Symbols

- Rituals, often accompanied by sacred symbols, serve as powerful tools for accessing divine wisdom. Lighting candles, chanting sacred words, or participating in communal ceremonies creates an atmosphere where the divine and human realms intersect.

- Symbols like the Tree of Life or the Flower of Life, which represent interconnectedness and divine order, act as visual reminders of the spiritual dimensions of reality.

4. Service to Others
- Enlightenment is deeply tied to selflessness and compassion. By serving others and acting in love, individuals embody divine wisdom in action. This principle is echoed in many traditions, where acts of kindness and justice are seen as reflections of divine will.

5. Engaging with Nature
- Nature serves as a manifestation of divine wisdom, with its rhythms, cycles, and patterns offering profound lessons. Observing the interconnectedness of ecosystems or the majesty of celestial movements can inspire awe and deepen one's connection to the divine.
- Many spiritual traditions encourage practices like walking in nature, contemplating its beauty, and aligning with its cycles as pathways to enlightenment.

6. Inner Transformation
- Enlightenment requires personal transformation letting go of ego, negativity, and attachments to the material world. By embracing humility, gratitude, and love, individuals open themselves to divine wisdom and the deeper truths of existence.

Role of Prophecy and Revelation

Prophecy and revelation are key ways through which divine wisdom has been communicated to humanity throughout history. These phenomena, often delivered through chosen individuals, serve as a means for the divine to guide, warn, and enlighten.

1. Prophecy as Divine Communication
- Prophets are seen as intermediaries between God and humanity. In the Book of Jubilees, figures like Enoch, Noah, and Abraham are portrayed as recipients of divine messages that reveal God's plans and expectations.
- Prophecies often address both immediate concerns and long-term visions, offering insights into human destiny, cosmic events, and the fulfillment of divine promises.

2. The Purpose of Revelation
- Revelation serves several purposes, including the unveiling of divine truths, guidance for living in alignment with divine laws, and assurance of God's presence and plan.
- In many cases, revelations are given during times of crisis or transition, providing hope and direction. For example, Moses' reception of the Torah on Mount Sinai represents a foundational moment of revelation, establishing the covenant between God and the Israelites.

3. The Nature of Prophetic Visions
- Prophetic visions often transcend ordinary perception, involving symbolic imagery, celestial beings, and messages from the divine. These visions are not confined to time and space, allowing prophets to see

past, present, and future events as part of a unified divine plan.

- In the Book of Jubilees, the concept of the "heavenly tablets" reflects this timeless nature, suggesting that all events are inscribed in a cosmic record.

4. Angelic Mediation in Prophecy

- Angels frequently play a role in delivering revelations. In the Book of Jubilees, the "angel of the presence" provides Moses with a detailed account of history and divine laws, emphasizing the connection between the heavenly and earthly realms.

- These angelic messengers highlight the importance of maintaining a spiritual connection to the divine and adhering to the revealed truths.

5. The Power of Symbolism in Revelation

- Prophetic revelations often use symbolic language and imagery to convey deeper truths. For example, the flood narrative in the Book of Jubilees symbolizes both divine judgment and the possibility of renewal, reflecting the cyclical nature of time and God's mercy.

- Understanding these symbols requires spiritual discernment, encouraging individuals to seek divine wisdom in their own lives.

6. Revelation as a Call to Action

- Prophetic messages are not merely informative; they are calls to action. They challenge individuals and communities to repent, change their ways, and align themselves with divine purpose.

- For instance, the Book of Jubilees calls for the observance of the Sabbath and sacred times as acts of obedience to divine law, reinforcing humanity's role in maintaining cosmic harmony.

The Connection Between Divine Wisdom, Prophecy, and Humanity

The interplay between divine wisdom and humanity is a two-way relationship. While humans seek enlightenment and understanding, divine wisdom is also actively revealed through prophecy and spiritual experiences. This dynamic connection fosters growth, transformation, and a deeper alignment with the divine plan.

1. Unity with the Divine
- By embracing divine wisdom, individuals align their actions and intentions with the cosmic order, creating harmony between their lives and the greater universe.
- This unity brings a sense of peace, purpose, and fulfillment, as humans recognize their place in the divine purpose of creation.

2. The Eternal Quest for Truth
- The human connection to divine wisdom is an ongoing journey. Each revelation or insight is a step closer to understanding the mysteries of existence and the mind of the Creator.
- This quest encourages humility, as individuals realize that divine wisdom is infinite and beyond full comprehension.

3. Living as Vessels of Wisdom
- Those who connect with divine wisdom are called to embody it in their lives, becoming vessels through which this sacred knowledge flows into the world. Through their actions, they reflect divine love, justice, and truth, inspiring others to seek their own connection.

The pursuit of divine wisdom and the human connection to it are central to understanding the universe's spiritual fabric. Pathways to enlightenment guide individuals toward this wisdom, while prophecy and revelation serve as divine interventions that illuminate the way. Together, they offer profound insights into the nature of existence, the purpose of life, and humanity's sacred relationship with the Creator

Part V
Reflections on Time and Redemption

Chapter 9

Jubilee Cycles and Human Evolution

The concept of jubilee cycles—a recurring pattern of 49 years divided into seven sabbatical cycles—offers profound insights into the nature of time, redemption, and human evolution. Rooted in ancient traditions, these cycles reflect the divine design of time as a sacred rhythm that governs both individual lives and collective histories. Each cycle serves as a spiritual reset, an opportunity for growth, renewal, and alignment with divine purpose.

Lessons from History

Throughout history, jubilee cycles have symbolized pivotal moments of redemption and transformation. These moments are not random but reflect recurring patterns that align with divine rhythms. By examining these patterns, humanity can uncover timeless lessons about its relationship with time, justice, and spiritual growth.

1. Periods of Liberation and Renewal
   - The jubilee, as described in Leviticus 25, mandated the release of slaves, the forgiveness of debts, and the return of ancestral lands. These practices ensured that

societal imbalances were periodically corrected, preventing the concentration of wealth and power while fostering equality and justice.

- Historically, these principles influenced the broader socio-political structures of ancient Israel, emphasizing the importance of cyclical renewal in maintaining harmony and stability.

2. Redemption Through Rest

- The sabbatical years within each jubilee cycle offered the land a period of rest, symbolizing humanity's dependence on divine providence rather than relentless toil. This principle underscores the importance of sustainable practices and the recognition of time as a divine gift.

- Modern ecological crises, such as soil degradation and climate change, echo the consequences of neglecting these sacred rhythms. Humanity's failure to honor the cycles of rest and renewal highlights the ongoing relevance of jubilee principles.

3. Moments of Transformation in Human History

- Significant historical events often align with jubilee-like patterns, suggesting that these cycles operate on both spiritual and historical levels. For example:

    - The emancipation of slaves in various cultures reflects the jubilee principle of liberation.
    - Economic reforms, such as debt forgiveness programs, mirror the ancient practice of resetting societal structures.

- These transformative moments demonstrate humanity's capacity to realign with divine justice and embrace collective redemption.

4. The Role of Forgiveness and Restoration
- The jubilee cycle emphasizes forgiveness—not just of debts but also of transgressions. This lesson extends beyond financial systems to interpersonal relationships and societal dynamics.
- Forgiveness, as a spiritual practice, enables individuals and communities to break free from cycles of resentment and conflict, fostering healing and reconciliation.

5. Patterns of Decline and Renewal
- History reveals recurring patterns of moral decline followed by periods of renewal. These cycles often coincide with spiritual awakenings or movements that call humanity back to divine principles.
- By recognizing these patterns, individuals and societies can anticipate periods of spiritual renewal and actively participate in them.

Prophetic Insights into Future Cycles

Prophecy, deeply intertwined with the concept of jubilee cycles, offers a forward-looking perspective on humanity's spiritual and societal evolution. These insights reveal how future cycles will unfold and what lessons humanity must embrace to align with divine will.

1. The Continuity of Divine Patterns

- Prophetic texts often emphasize that the cycles established by God are eternal, extending into the future as markers of divine order. The Book of Jubilees presents time as a preordained sequence, recorded on heavenly tablets and governed by cosmic rhythms.

- This continuity suggests that jubilee principles—liberation, forgiveness, and renewal—will remain central to humanity's evolution, guiding it toward ultimate redemption.

2. Signs of the Times

- Prophecies frequently highlight specific signs that indicate the approach of significant jubilee-like events. These signs, such as widespread injustice, ecological disruptions, and spiritual decline, call for collective repentance and renewal.

- Modern challenges, including economic inequality and environmental degradation, can be interpreted as signals of humanity's need to realign with the principles of jubilee cycles.

3. The Promise of Ultimate Redemption

- Many prophetic texts, including those influenced by jubilee principles, point toward a final redemption—a time when creation will be fully restored to its intended harmony. This ultimate jubilee will transcend historical cycles, ushering in a new era of peace, justice, and divine presence.

- This vision of redemption inspires hope, encouraging humanity to persevere through challenges and actively participate in the unfolding of God's plan.

4. The Role of Human Agency in Future Cycles
- Prophecies often emphasize the importance of human agency in shaping future cycles. While divine patterns are immutable, humanity's response to these patterns determines the nature of its experiences.
- By embracing forgiveness, justice, and sustainability, individuals and societies can align with divine rhythms, ensuring that future cycles are marked by renewal rather than judgment.

5. The Interplay of Time and Eternity
- Prophetic insights reveal that time is not merely linear but a sacred spiral that bridges the temporal and eternal. Each jubilee cycle is a step closer to ultimate redemption, drawing humanity deeper into the mysteries of divine purpose.
- Understanding this interplay encourages individuals to view their lives as part of a greater cosmic journey, fostering patience, faith, and a commitment to spiritual growth.

The Spiritual Significance of Jubilee Cycles for Human Evolution

The concept of jubilee cycles offers profound insights into humanity's spiritual and societal evolution. These cycles serve as reminders of humanity's dependence on divine wisdom and its capacity for renewal. By embracing the lessons of the past and heeding prophetic insights into the future, individuals and societies can align with divine rhythms, fostering harmony and redemption.

1. A Call to Align with Sacred Time
- Jubilee cycles remind humanity that time is sacred and must be respected. Aligning with these cycles fosters spiritual growth and ensures sustainable practices that honor both creation and the Creator.

2. Opportunities for Collective Redemption
- Each cycle offers a chance for societies to correct imbalances, address injustices, and embrace forgiveness. These opportunities are not just historical but deeply spiritual, reflecting God's mercy and desire for restoration.

3. Hope for the Future
- The prophetic promise of an ultimate jubilee—a time of eternal peace and restoration—offers hope and inspiration. It reminds humanity that the challenges of the present are part of a greater journey toward divine fulfillment.

4. Lessons for Personal and Collective Growth
- On an individual level, jubilee cycles teach the importance of rest, renewal, and forgiveness. On a collective level, they emphasize justice, equality, and

sustainability. Together, these lessons guide humanity toward a more harmonious and enlightened existence.

The jubilee cycles, as reflections of divine order, are integral to understanding human evolution and the unfolding of God's plan. By studying these cycles, humanity can gain insights into its past, present, and future, embracing the principles of redemption and renewal that lie at the heart of creation. Through alignment with these sacred rhythms, individuals and societies can participate in the ongoing process of spiritual transformation, moving ever closer to ultimate redemption.

Chapter 10

Redemption Through Sacred Knowledge

Redemption through sacred knowledge is a timeless principle that underscores humanity's journey toward spiritual awakening and restoration. Sacred knowledge is not merely intellectual but deeply transformative, encompassing divine truths, mystical insights, and eternal wisdom that guide individuals toward harmony with the universe and their Creator. This redemptive process involves understanding and applying sacred knowledge to transcend the limitations of the material world, heal from spiritual disconnection, and align with divine purpose.

Embracing Mystical Teachings

Mystical teachings form the foundation of sacred knowledge, offering profound insights into the nature of existence, the divine, and the human spirit. These teachings, often encoded in symbols, parables, and ancient texts, invite individuals to look beyond the surface of reality and explore the deeper, spiritual dimensions of life.

1. The Transformative Power of Mystical Teachings

 - Mystical teachings emphasize inner transformation as the key to redemption. They encourage individuals to look inward, confront their limitations, and embrace spiritual practices that awaken divine wisdom within.

- Traditions such as Kabbalah in Judaism, Sufism in Islam, and Christian mysticism all focus on the inner journey, offering pathways to self-discovery and unity with the divine.

2. Sacred Geometry and the Language of Creation

- Sacred geometry, often referred to as the blueprint of creation, reveals the divine patterns and structures that underpin the universe. Shapes like the Flower of Life and the Tree of Life symbolize interconnectedness and divine order.

- By contemplating these patterns, individuals gain a greater appreciation for the harmony of creation and their place within it, fostering a deeper connection to the divine.

3. The Mystical Nature of Sacred Texts

- Sacred texts such as the Book of Jubilees, the Torah, and the Vedas contain layers of meaning that go beyond their literal interpretations. Mystical traditions encourage readers to seek the hidden truths within these texts, uncovering divine wisdom that can guide their lives.

- For instance, the Book of Jubilees presents history as a series of divine cycles, emphasizing the importance of aligning human actions with cosmic rhythms.

4. The Role of Rituals and Symbols

- Mystical teachings often employ rituals and symbols to bridge the gap between the material and spiritual

worlds. Rituals like meditation, chanting, and prayer open channels for divine energy, while symbols such as the cross or the mandala serve as focal points for spiritual reflection.
   - These practices help individuals internalize sacred knowledge, transforming it from abstract understanding into lived experience.

5. Unity and Oneness
   - A central theme of mystical teachings is the interconnectedness of all things. By embracing the idea of oneness, individuals transcend divisions and realize that their actions, thoughts, and intentions affect the entire universe.
   - This realization fosters compassion, humility, and a sense of responsibility for the well-being of others and the world.

Practical Applications for Modern Times

In an era defined by rapid technological advancement and material pursuits, applying sacred knowledge to modern life is more essential than ever. Mystical teachings provide practical tools for navigating the complexities of contemporary existence while fostering spiritual growth and alignment with divine principles.

1. Mindfulness and Presence
   - Sacred knowledge teaches the importance of living in the present moment, free from distractions and anxiety about the future. Practices like mindfulness

meditation, inspired by mystical traditions, help individuals cultivate awareness and connect with their inner selves.

- In modern times, mindfulness can reduce stress, enhance focus, and promote emotional resilience, creating a foundation for spiritual awakening.

2. Sustainability and Stewardship

- The principle of interconnectedness emphasized in sacred knowledge calls for greater responsibility toward the environment. By recognizing the divine in all creation, individuals are inspired to adopt sustainable practices that honor the Earth as a sacred gift.

- This includes actions such as reducing waste, conserving resources, and supporting ethical businesses, aligning daily choices with spiritual values.

3. Ethical Living and Social Justice

- Mystical teachings stress the importance of living ethically, with compassion and integrity. In modern society, this translates into advocating for social justice, treating others with kindness, and working to eliminate inequality and suffering.

- Sacred knowledge reminds individuals that their actions have far-reaching consequences, encouraging them to be agents of positive change.

4. Healing and Personal Growth

- Sacred knowledge offers tools for personal healing and growth, addressing emotional wounds and spiritual disconnection. Practices like journaling, energy healing, and affirmations draw from mystical traditions to help individuals release negativity and embrace their true potential.

- These practices are particularly valuable in modern times, where many struggle with feelings of isolation and disconnection.

5. Technology as a Tool for Connection

- While modern technology can be a source of distraction, it can also serve as a tool for spreading sacred knowledge. Online communities, meditation apps, and digital libraries make mystical teachings accessible to people around the world.

- By using technology mindfully, individuals can deepen their spiritual practice and connect with others who share similar values.

6. Building Sacred Communities

- Sacred knowledge emphasizes the power of community in fostering spiritual growth. In modern times, individuals can create or join communities that support shared values, such as meditation groups, book clubs focused on spiritual texts, or sustainability initiatives.

- These communities provide support, inspiration, and accountability, helping members stay aligned with their spiritual goals.

7. Balancing the Material and Spiritual

- One of the greatest challenges of modern life is balancing material success with spiritual fulfillment. Sacred knowledge teaches that material wealth is not inherently negative but must be used in service of higher purposes.

- By aligning material pursuits with spiritual values, individuals can create a life that is both prosperous and meaningful.

The Promise of Redemption Through Sacred Knowledge

Sacred knowledge offers a path to redemption by transforming how individuals perceive themselves, others, and the universe. It provides a framework for understanding the divine order of creation and offers tools for aligning human actions with this order. Redemption, in this sense, is not just about personal salvation but about contributing to the collective evolution of humanity.

1. Healing Division

- By embracing the principles of unity and oneness, sacred knowledge helps heal divisions between people, cultures, and nations. It reminds humanity of its shared destiny and the importance of working together for the greater good.

2. Cultivating Inner Peace

- Sacred knowledge fosters inner peace by encouraging practices that align individuals with divine rhythms. This peace radiates outward, influencing relationships, communities, and the world.

3. Fulfilling Divine Purpose

- Redemption through sacred knowledge involves recognizing and fulfilling one's divine purpose. This purpose is unique to each individual but ultimately contributes to the greater harmony of creation.

Sacred knowledge is a timeless gift that holds the key to redemption and spiritual fulfillment. By embracing mystical teachings and applying them to modern life, individuals can navigate the challenges of the contemporary world while remaining connected to divine wisdom. This journey toward redemption is not only personal but collective, offering humanity the opportunity to evolve in harmony with the sacred rhythms of the universe.M

Chapter 11

Timeline of Jubilees and Significant Events

The timeline of jubilees and significant events provides a structured overview of pivotal moments in history where the principles of jubilee—restoration, liberation, and renewal—manifested in human and divine narratives. This timeline is not merely a chronological listing but a reflection of humanity's evolving relationship with the divine, guided by sacred rhythms and cosmic cycles.

1. The Beginning of Time and Creation
- The Creation Week (Genesis 1-2): The foundation of sacred time begins with the creation of the universe in six days, culminating in the seventh day of rest, which serves as the prototype for the Sabbath and the jubilee cycle.
- Establishment of Sacred Rhythms: God sets the celestial bodies sun, moon, and stars—as markers of days, months, and years, embedding the principle of time into creation itself.

2. The Patriarchal Period
- Adam and Eve (Year 1): The first humans experience the divine rhythm of life and the consequences of

disobedience, setting the stage for humanity's journey toward redemption.
- Enoch's Walk with God (Year 987): Enoch, the seventh generation from Adam, embodies alignment with divine will and is taken by God, signifying the first significant fulfillment of divine-human connection in a jubilee context.
- Noah and the Flood (Year 1656): The flood marks a reset of creation, reflecting the jubilee principle of cleansing and renewal, with Noah's family symbolizing a new beginning for humanity.

3. The Covenant with Abraham
- Abraham's Covenant (Year 1948): God's promise to Abraham introduces the concept of land restoration and spiritual inheritance, central themes in the jubilee framework.
- Birth of Isaac (Year 2048): Isaac's birth signifies the fulfillment of divine promise, aligning with the themes of divine timing and redemption.

4. The Exodus and the Giving of the Law
- The Exodus from Egypt (Year 2454): The liberation of the Israelites from slavery is a historical enactment of jubilee principles—freedom from bondage, restoration of identity, and renewal of purpose.
- Mount Sinai and the Torah (Year 2455): The giving of the Torah, including the laws of the sabbatical year and jubilee, establishes the sacred structure of time and justice for the Israelites.

5. The Establishment of Israel
- Entry into the Promised Land (Year 2488): The division of land among the tribes of Israel reflects the jubilee principle of land inheritance and equitable distribution.
- First Observance of the Jubilee (50th Year in the Land): Upon settling in the Promised Land, the Israelites are commanded to observe the first jubilee year, reinforcing the cycle of restoration and renewal.

6. The Period of Kings and Prophets
- King David's Reign (Year 2891): David's establishment of Jerusalem as the spiritual and political center signifies a moment of renewal and divine favor, aligning with jubilee themes of restoration.
- Prophets and Calls for Repentance: Prophets such as Isaiah and Jeremiah echo jubilee principles, calling for justice, liberation, and alignment with God's covenant.

7. The Babylonian Exile and Return
- Destruction of Jerusalem (Year 3358): The fall of Jerusalem and the Babylonian exile reflect a disruption in sacred rhythms, emphasizing the consequences of ignoring jubilee principles.
- Return from Exile (Year 3428): Under Persian King Cyrus, the Israelites are allowed to return to their land, marking a significant act of restoration and fulfillment of jubilee themes.

8. The Life and Ministry of Jesus Christ

- Birth of Jesus (Year 3967): Jesus' arrival represents the ultimate jubilee, offering redemption and liberation for all humanity.
- Proclamation of Jubilee (Luke 4:16-21): In His inaugural sermon, Jesus declares the fulfillment of the jubilee, emphasizing freedom for the oppressed, healing, and restoration.
- Crucifixion and Resurrection (Year 4000): Jesus' death and resurrection embody the ultimate act of redemption, aligning with the cosmic jubilee that reconciles humanity with God.

9. The Early Church and Global Impact

- Pentecost (Year 4000+50 Days): The descent of the Holy Spirit marks the establishment of the church, symbolizing a new era of spiritual liberation and renewal.
- Spread of the Gospel: The teachings of Jesus as the fulfillment of the jubilee cycle spread globally, emphasizing forgiveness, grace, and restoration.

10. The Modern Era and Future Jubilees

- Abolition of Slavery (19th Century): The global movement to abolish slavery reflects the jubilee principle of liberation from bondage and the restoration of human dignity.
- Economic Resets and Debt Forgiveness: Initiatives such as debt forgiveness for developing nations echo ancient jubilee practices, emphasizing justice and equity.

- Environmental Movements: Modern efforts to restore balance to the environment align with the jubilee principle of letting the land rest and practicing sustainable stewardship.

11. The Final Jubilee and Ultimate Redemption
- Prophetic Fulfillment: The ultimate jubilee, as foretold in prophetic texts, will culminate in the restoration of all creation, the defeat of evil, and the establishment of eternal harmony.
- New Heaven and New Earth (Revelation 21-22): The final act of redemption, described in Revelation, reflects the completion of the cosmic jubilee, where time itself is fulfilled, and divine order is fully restored.

Reflection on the Timeline

This timeline of jubilees and significant events highlights humanity's ongoing journey through cycles of redemption, renewal, and restoration. It underscores the interconnectedness of divine principles with historical and cosmic rhythms, inviting reflection on the relevance of jubilee teachings in every age. By understanding and honoring these cycles, humanity participates in the unfolding of God's eternal plan, moving ever closer to the ultimate jubilee where all things are made new.

Conclusion

The Eternal Quest for Understanding

Throughout history, humanity has pursued an unending quest to understand the universe, its origins, and the intricate forces that govern existence. This quest has been fueled by a profound desire to connect with something greater—a divine intelligence that orchestrates the rhythms of creation and provides meaning to the mysteries of life. The book of Jubilees explores this eternal pursuit, weaving together ancient wisdom, sacred teachings, and cosmic insights to illuminate the interconnectedness of time, space, and spiritual purpose.

The jubilee, as both a concept and a practice, serves as a metaphor for this quest. It reflects humanity's deep-seated need for periodic renewal, restoration, and alignment with divine order. By viewing time as sacred and cyclical, the jubilee offers a framework for understanding the human journey—not as a linear path but as a series of transformative cycles that mirror the divine rhythms of the universe. Each jubilee cycle becomes an opportunity to reconcile with the Creator, to forgive and be forgiven, and to realign with the higher purpose of existence.

In seeking to understand the jubilee mysteries, we uncover profound truths about the nature of redemption, the interplay of the celestial and terrestrial, and humanity's role in the grand design of the cosmos. These truths are not confined to ancient traditions but resonate deeply with modern challenges, offering timeless wisdom for navigating the complexities of contemporary life.

Closing Thoughts on the Jubilee Mysteries

The book of Jubilees: Unveiling Mystical Dimensions of the Universe serves as both a guide and an invitation. It guides readers through the sacred rhythms of time, the mystical dimensions of the universe, and the profound connection between the human and the divine. It invites us to reflect on the enduring relevance of the jubilee mysteries and their capacity to transform not only individual lives but entire societies.

The jubilee reminds us of the importance of balance, justice, and renewal. It calls for the restoration of harmony in our relationships, our communities, and the natural world. In a world often marked by disconnection and discord, the principles of the jubilee—liberation, forgiveness, and unity—offer a pathway to healing and hope. It challenges us to live not merely in accordance with human desires but in alignment with divine rhythms, ensuring that our actions contribute to the greater harmony of creation.

As we close this exploration of the jubilee mysteries, we are reminded that the quest for understanding is not an end but a journey—a perpetual cycle of seeking, discovering, and transforming. The lessons of the jubilee encourage us to embrace this journey with humility, curiosity, and reverence. They teach us that redemption is not a distant goal but an ever-present possibility, unfolding in the sacred rhythms of time and the choices we make each day.

The mysteries of the jubilee are, in essence, the mysteries of life itself—woven into the fabric of existence, waiting to be discovered and lived. By embracing these mysteries, we not only gain a deeper understanding of the universe but also fulfill our sacred calling as stewards of creation and participants in the divine dance of time and eternity.

Let the principles of the jubilee guide us toward a future marked by harmony, redemption, and a renewed connection to the divine. For in understanding the jubilee, we come closer to understanding ourselves, our Creator, and the eternal truths that unite all things.

www.ingramcontent.com/pod-product-compliance
Lightning Source LLC
Chambersburg PA
CBHW071653240526
45469CB00023B/2366